Is It Alive?

Marcia S. Freeman

Contents

Some things on Earth are alive.
All the other things on Earth are not.

How can you tell what is alive?

Brown Booby

3

All living things grow and change.
You are growing right now, but you
grow so slowly that you hardly notice.

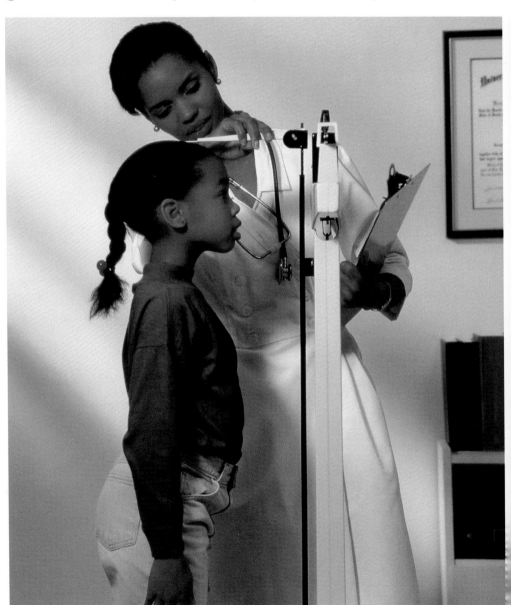

Some things, like these bulbs, grow quickly. You can see the change in a few days.

Day 1   Day 4   Day 14

These rocks do not grow.
They are not alive.
Nonliving things do not grow.

Living things **reproduce**. Plants make seeds or grow shoots. Animals lay eggs or have live babies.

butterfly laying eggs

Which of these things
are alive? Which are not?
How do you know?

All living things need food to **survive**. Plants make their own food from sunlight, air, and water.

Animals eat plants or other animals.

Living things need water.

Many animals drink fresh water
every day.

Most plants **absorb** water through their roots.

Living things make wastes. Animals **excrete** wastes from their bodies.

Many things can move from place to place. Some move quickly. Others move slowly.

Which of these things are moving on their own? What is making the other things move?

Some nonliving things were
never alive.

These bottles are made of
glass, which was never alive.

Other nonliving things were once
alive. This cabin is made of logs.
The logs came from living trees.

These leaves were
alive before they
fell from a tree.

All living things are made up of living parts called **cells**. Cells are like tiny building blocks.

They are so small that you need a **microscope** to see them.

microscope

lilac leaf cells

If something grows, reproduces, uses food and water, and moves on its own—you can be sure it is alive.

Which of these things is alive?
How can you tell?

Glossary

absorb (uhb-ZOHRB): take in; soak up

cells (SELZ): the tiny units that make up all living things and things that once lived

excrete (eks-KREET): get rid of waste from the body

microscope (MYE-kruh-skohp): an instrument that makes small things look large and helps people see tiny things such as cells

nonliving (NON-LI-ving): not alive

reproduce (ree-pruh-DOOS): make another; produce seeds, eggs, or babies

survive (sur-VYV): stay alive